Adolph Schoenman

Milk Testing

Instructions for testing milk and dividing money for creameries, cheese

factories and dairymen

Adolph Schoenman

Milk Testing
Instructions for testing milk and dividing money for creameries, cheese factories and dairymen

ISBN/EAN: 9783744795883

Printed in Europe, USA, Canada, Australia, Japan

Cover: Foto ©berggeist007 / pixelio.de

More available books at **www.hansebooks.com**

HIRAM SMITH HALL – WISCONSIN DAIRY SCHOOL BUILDING.

MILK TESTING

INSTRUCTIONS FOR TESTING MILK AND
DIVIDING MONEY FOR

Creameries, Cheese Factories and Dairymen

BY

ADOLPH SCHOENMAN

INSTRUCTOR IN MILK TESTING, UNIVERSITY OF WISCONSIN

MADISON, WIS.
PUBLISHED BY THE AUTHOR
1894

Tracy, Gibbs & Co., Printers, Madison, Wis.

PREFACE.

It has for many years past been a recognized fact of all the leading Experimental Stations that a simple and accurate method of determining the butter fat of milk was sorely needed for the the general good of the Dairy public.

Although the general dairy public is rather slow in "catching on" to the great injustice of pooling milk by *weight only*, regardless of the fat it contains.

The wiser heads and experimental workers have for several years discovered the great injustice of the "weight pooling" practice. And further than that, they have seen the great mass of dairy farmers, year after year, feeding cows of all sorts, good, bad and indifferent, thousands of which were not paying for their keep, and are a curse, not a boon, to their owner.

But with no better method at hand than to cream and churn each cow's milk separate for the purpose of weeding out the poor ones, it would be needless to preach cow testing to the average dairy farmer.

Therefore, for the double reason as above stated, the invention of a simple and accurate devise for measuring the butter fat of milk was ardently sought for by the chemists of several of the leading Experimental Stations.

Several years ago Professor Short of the Wisconsin Experimental Station led the way by inventing a method by which the butter fat of milk could be quite readily measured. Although not quite satisfactory, it was a stride in the right direction. Next came the test of Professor Patrick of the Iowa Experimental Station, which was somewhat different and in a measure quite satisfactory. But not quite *the* thing for quick, simple, and accurate work. But to cap the climax, Dr. S. M. Babcock, chief chemist of the Wisconsin Experimental Station invented a simple and accurate test by which the average school boy of fourteen years of age, by carefully reading the instructions can make an accurate butter fat test of a dozen different cows in ten to fifteen minutes time.

The great and wonderful good this invention which Dr. Babcock gave to the dairy public *free* time alone can tell. Through the courtesy of Dr. Babcock, by whom the writer was greatly assisted in getting up this little book, the author takes this opportunity to thank the doctor for his kindness.

Part I gives reasons why the test should be applied to the cows for the owner's sake, and at the factory it should be applied for justice's sake.

Part II gives a complete description for making the test, not only for fat, but also for solids not fat, and for finding adulterations. It also gives a full and concisive description of the composite test, and, finally, shows by actual examples how to divide the money under the test.

Part III gives useful pointers which are indexed and numbered in a manner which can not fail to simplify the subject and increase the usefulnes of the book. The writer believes, however, that to thoroughly master the subject the student should not only study each pointer separately, but should reinforce his knowledge in each case by an actual experiment.

                                        THE AUTHOR.

Plain, Wis.

TABLE OF CONTENTS.

PART I.

PART II.

PART III.

PART I.

APPLICATION OF THE TEST.

1. Reasons why Dairymen Should Apply the Babcock Test to Their Cows.

Farmer Jones is the owner of three cows; his favorite cow is Bess, a fine large cow which gives a large yield of milk. While Bell is a puny looking animal with only a moderate milk yield. "Daisy," he says, "will have to be sold. She gives only about three fourths as much milk as Bess and eats just as much food." He had formed an opinion of each cow, judging only from quantity (as most dairymen do), while quality was not considered.

The cheese maker (who owned a test), had on several occasions heard of Farmer Jones' Bess and her large milk yields. So, one fine day, he went down to test her, and also the other two, and obtained the following result:

Butter Fat.

Bess — Daily yield, 32 pounds milk. Test, 2.8 % = .896 pounds.
Bell —   "   "   25   "   "   "   4.0 % = 1.00   "
Daisy—   "   "   22   "   "   "   4.6 % = 1.012   "

Farmer Jones: Here is the result of the test of your three cows. Daisy is your best cow, yielding 1.012 pounds of butter fat per day.

Bell comes next, with a record of one pound of butter fat, while Bess, your brag cow, brings up the rear with a record of .896 pounds of butter fat.

"I am astonished at those results, Mr. Cheese-maker, and now see that through my ignorance of judging a cow by the quantity of her milk, regardless of quality, I came nearly selling my best cow at a cow-beef price, and now this little Babcock machine told me in a ten minutes' test that she is a jewel indeed, and is not for sale at any price.

Isn't that a dandy little machine, though? But I suppose, of course, it is patented and costs a pile of money."

"No, sir; it is not patented. Dr. Babcock gave this wonderful invention to the dairy public as free as the water that flows from the well."

Dear reader, the foregoing is simply a correct illustration of a false notion of a dairy cow based on the deceitful and misleading basis of *quantity* alone. And the wonderful results wrought by applying the Babcock test to a herd of cows. When we think of the thousands of herds of cows throughout the land which are kept on the basis of *quantity* and the large percentage of which do not pay for their keep, and are further acquainted with the fact that any farmer can now purchase a four-bottle Babcock test at the nominal price of $5.00, wherewith, in connection with a pair of spring balance he can purify his herd by weeding out the unprofitable portion thereof with wonderful accuracy and great profit. Yes; when we think of all those things, we wonder in amazement of the future greatness of the Babcock test.

Look at these figures.

Here are the tests of six cows kept at the Wisconsin Experimental Farm, and probably fed and cared for exactly alike:

Milk of Bessie   tested 6.95 per cent. of fat.
"   " Jersey   "   6.91   "   "
"   " Sylvia   "   6.44   "   "
"   " Mattie   "   3.28   "   "
"   " Bunn   "   2.87   "   "
"   " Topsy   "   3.35   "   "

The average of the first three is 6.76, while the average of the last three is only 3.16. Suppose the milk of the former is worth $1.00 per hundred, the latter is worth less than fifty cents.

2. Why Cheese Factories Should pay by the Test.

The cheese of our factories will never attain a high standard nor a high price, as long as the method of pooling milk by weight only, prevails. By this method the cheese factory patrons are constantly struggling to deliver weight. Weight is money, and the farmer schemes and studies how to deliver a large amount, *in weight*, either by honest or dishonest means.

Since the short advent of the Babcock test it has been proven by many Experiment Stations and otherwise, a hundred fold, that the value of cheese up to $4\frac{1}{2}$ per cent. or 5 per cent. milk as a rule corresponds to the amount of butter fat it contains.

Every thinking man will at once see that the pooling of milk by weight only, offers a premium on poor milk and thereby degrades the milk standard to a low level.

"Why," Mr. A. says, "I breed a strain of cows that yield a large *quantity*, my neighbor B., who is not so shrewd as I, can furnish the *quality*." Quantity is money in pooling milk by weight only, and the shrewdest men willingly degrade the milk to the thinnest of thin milks and thereby degenerate our cheese to a cheese of poor quality and poor price.

While on the other hand if cheese factories pay by the Babcock test they offer a premium on a thing of merit, viz: On good rich milk. The idea now is: "The more butter fat the more money."

Now Mr. A. will squirm and kick and squeal. But there is no hope for him. The test plan is bound to win in the end. And why?

Because in the test plan the premium is offered to the man that brings the most butter fat. *A thing of merit and of worth.*

The writer has made cheese for many years, having taken weekly tests and found that on an average the richest milk has invariably tested one per cent. above the thinnest, (all the cows being common native cows). Considering butter fat worth twenty-four cents we find that C., the man with the richest milk, has invariably furnished butter fat to grease the man's cheese which brought the thinnest milk to the tune of twelve cents per cwt. of milk, to bring them up to the average standard of our factory cheese.

Suppose each of these men furnished 85,000 pounds of milk per season. We find that C. has furnished butter fat to the snug little sum of $102.00 to grease his neighbor's cheese.

These are undisputable facts, and we hope every cheese maker and milk producer will give them a careful study. The reader will readily see that in factories where Jerseys and Guernseys are mixed with common cows the injustice might be much more marked.

There is another reason that will greatly tend to bring the cheese factories to pay by the test plan, where creameries and cheese factories are intermixed. And that is: Creameries will pay by the test, and draw all the rich milk to their doors, and factories will be compelled to follow the creameries' example or work at a great disadvantage.

3. Useful Hints to Cow Owners.

I. Buy a Babcock test and find out the real worth of your cows as "cheese maker" did for Farmer Jones and saved the life of a precious animal.

II. A four-bottle tester costs $5.00 and if rightly applied to a herd of miscellaneous unknown cows (a farmer never knows a cow until tested) with the object in view of "weeding out" and "grading up," it may bring in, in a short time, its cost in a hundred fold.

III. Butter fat is money and the Babcock test will tell you where to find it. It will point you out which cows to keep and which to sell in ten minutes time.

IV. The Babcock test tells the story of the terrible loss of butter fat by the average gravity method. It has shown that the average loss in gravity creaming is about three-fourths pounds per 100 pounds of milk. While in centrifugal creaming it is about one-sixth pound only.

V. Did you ever dream that a pound of butter fat is worth about one hundred and fifty times as much as a pound of skim milk?

VI. If you did, it must have dawned upon you by this time that it is a gross injustice to buy and sell them at the same price; which is done by the "weight only method" of pooling milk.

VII. Suppose you have ninety-seven pounds skim milk and three pounds of butter fat and your neighbor has ninety-six pounds skim milk and four pounds of butter fat. By the "weight only method" of pooling milk you both get the same price. Nevertheless 300 pounds of your neighbor's milk is worth as much as 400 pounds of your milk. And now you kick and squirm and grumble because the Babcock test has revealed this terrible injustice, and does not let you continue the draught on your neighbor's pocket book.

VIII. The writer has just completed a test from a a sample of buttermilk taken from a farmer's churn, and it tested 3.2 per cent. fat. At the rate of sixteen and two-thirds pounds buttermilk per cwt. of milk (which is the rule under the Cooley system), this farmer keeps every sixth cow for waste in buttermilk· The Babcock test has told him this terrible story of waste, and he is now studying ways and means to stop the leak, and save the product of that sixth cow. Amen.

IX. The great worth of the Babcock test is in the following items: First—it reveals the great losses in setting milk in the usual manner as practiced by most dairymen. Second — It points out the great losses which occur in churning thin cream at a high temperature. Third—It is the best cow herd purifier known to man. Fourth—It has shown and continues to show the terrible injustice of "weight only" pooling system.

PART II.

THE GLASSWARE AND MACHINERY OF THE BABCOCK TEST.

4. The Regular Bottles. [Fig. 1.] The regular Babcock test bottle should contain at least 40 c. c. up to the neck. The neck is graduated from 0 to 10 per cent. Each division of the graduated scale represents .04 c. c. Five of those divisions are equivalent to one per cent. of fat, when one pipette of 17.6 c. c. milk is used.

5. The Pipette. [Fig. 2.] The pipette should contain, when filled to the mark, 17.6 c. c. A pipette of this size will deliver a little less than 17.5 c. c. and when of milk of average specific gravity, will weigh 18 grams. The pipette should be accurately calebrated. It can be tested by weighing the amount of mercury necessary to fill it to the mark. The weight of mercury should be 239 grams. Always be sure and buy a pipette marked 17.6 c. c. There are other sized pipettes on the market but they are "fool" pipettes and should never be used.

6. Acid Measure. [Fig. 3.] A glass cylinder with a lip to pour from and a single mark at 17.5 c. c. is the best form for general use.

7. Cream Bottles [Fig. 4.] are the same as the regular bottle except that they have a bulb in the neck capable of holding 10 per cent. of fat.

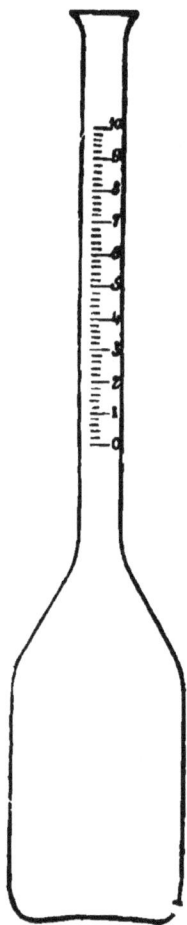
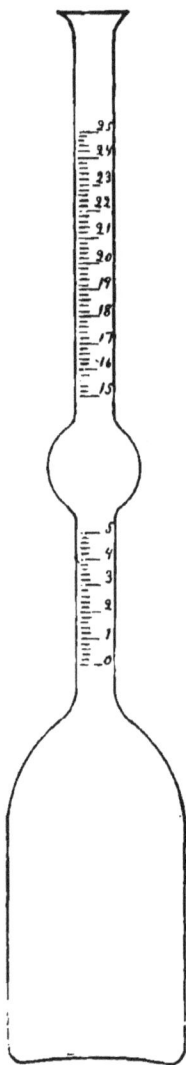

FIG. 1.   FIG. 4.   FIG. 2.   FIG. 3.

8. Skim Milk Bottles are capable of holding twice the amount of the regular bottle, and when they are used it should be remembered that two pipettes of milk and two measures of acid are delivered. Each division on the scale of the neck on this bottle is equivalent to one-tenth per cent. of fat, instead of two-tenths per cent. as is the case in the regular bottle where only one measure of milk and one of acid is used.

9. Machine for Whirling. There are many different styles of machines, but all operating on the same principle. A machine should be capable of making from 700 to 1,200 revolutions per minute. A small wheel should make more revolutions than a large one.

10. About the Motion. In machines where the motion is transmitted by belt or friction, the adjustment should be kept tight enough to avoid slipping, as otherwise the motion may be much less than is intended, and result in an imperfect separation of the fat.

11. The Acid. Commercial sulphuric acid having a specific gravity of 1.82 to 1.83 should be used. The stronger is preferable. It is very important that the acid used have approximately the right strength. If it is considerably too weak the casein will not all be cut out, and being mingled with the fat will give an unsatisfactory test.

If the acid is only a trifle too weak, the use of a little more may give a good test. If the acid is too strong it will turn the fat to a dark color. A good test may be obtained with too strong acid by using a little less acid. The acid should not be diluted.

12. Weak Acid. If acid is only a trifle too weak you will get good results by warming the milk to 70° or 75° each test. If your acid is so weak that when testing milk at 50° you have a white sediment in the lower end of the fat column, you may get good results by testing the milk at 70° or 75° with the same acid.

13. Boiling Water. Boiling water should be provided for filling the bottles after they have been whirled for the first time, and for warming the contents of the bottles in cool weather. Distilled or rain water is the best for filling the bottles.

MAKING THE TEST.

14. Sampling the Milk. Every precaution should be taken to have the sample represent as nearly as possible the whole lot of milk from which it is taken.

Milk fresh from the cow while still warm and before the cream is separated in a layer may be thoroughly mixed by pouring three or four times from one vessel to another. Milk that has stood until a layer of cream has formed should be poured more times, until the cream is thoroughly broken up and the whole appears homogeneous.

No clots of cream should appear upon the surface when the milk is left quiet for a moment. Milk should not be poured more times than is necessary, as extended mixing in this way is liable to churn the cream forming little granules that quickly rise to the surface. When this occurs it is impossible to obtain a fair sample and it is useless to make the test. Milk is sometimes churned by being transported long distances in vessels that are not full.

15. Measuring the Milk. When the milk has been sufficiently mixed the milk pipette is filled by placing its lower end in the milk can and sucking at the upper end until the milk rises above the mark on the stem; then remove the pipette from the mouth and quickly close the tube at the upper end by firmly pressing the end of the index finger upon it to prevent access of air. Then carefully relieve the pressure on the finger so as to admit air slowly to the space above the milk. Always have the upper end of the pipette and the finger dry when you measure milk, as it is almost impossible to gradually lower the milk with a wet finger. When the milk is lowered to the mark on the pipette press suddenly with the finger to prevent the milk from flowing out. Next place the point of the pipette in the mouth of one of the test bottles, held in a slightly inclined position so that the milk will flow down the side of the tube, and remove the finger allowing the milk to flow into the bottle. Then blow into the upper end to expel the drop of milk held in the point.

16. Adding the Acid. When all the samples of milk to be tested are measured ready for the test, the acid measure is filled to the 17.5 c. c. mark with sulphuric acid and from this it is carefully poured into a test bottle, containing milk, that is held in a slightly inclined position. The acid being much heavier than the milk sinks directly to the bottom of the test bottle without mixing with the milk that floats upon it. The acid and milk should be thoroughly mixed together by gently shaking with a rotary motion.

17. Whirling the Bottles. The test bottles containing the mixture of milk and acid should be placed in the machine and whirled directly after the acid is added and mixed. In even numbered testers an even number of bottles should be whirled at the same time, and they should be placed in the wheel in pairs, opposite each other, so that the equilibrium of the wheel will not be disturbed. The test should never be made without the cover on the jacket. After seeing that your bearing on the machine is all right, whirl the bottles at the proper speed about five minutes; then stop, and with pipette fill the bottles with hot distilled or rain water to about the 7 per cent. mark, replace the cover and whirl the bottles about one minute. Now you are ready to measure the fat in the neck of the bottle.

Never attempt to measure the fat on any test bottle unless it is warm enough for the fat to be quite liquid. If the temperature in the room is cool, be sure and have a pail of hot water at hand when the whirling is completed. Then immerse your bottles in hot water up to the figure ten on the neck, taking them out as you get ready to measure the fat. It is impossible to get a good reading unless the fat is quite hot.

18. Measuring the Fat. The fat when measured should be warm enough to flow readily, so that the line between the acid liquid and the column of fat will quickly assume a horizontal position when the bottle

is removed from the machine.   Any temperature between 110° F. and 150° F. will answer, but the higher
temperature is to be preferred.   The slight difference
in the volume of fat due to this difference in temperature is not sufficient to materially affect results.

To measure the fat, take a bottle from its socket,
and holding it in a perpendicular position with the
scale on a level with the eye, observe the divisions
which mark the highest and the
lowest limits of·the fat.   The difference between these gives the
per cent. of fat directly.   The
reading can easily be taken to half
divisions, or to one-tenth per cent.

The line of division between the
fat and the liquid beneath is nearly a straight line and no doubt
need arise concerning the reading
at this point, but the upper surface of the fat being concave, er-
rors often occur by reading from
the wrong place.   The reading
should be taken at the line where
the upper surface of the fat meets the side of the tube
and not from surface of fat in the center of the tube
nor from the bottom of the dark line caused by the
refraction of the curved surface.   For instance, in
Fig. 5 the reading should be taken from a to b and
not to c or d.

The reading may be made with less liability of error
by measuring the length of the column of fat with a

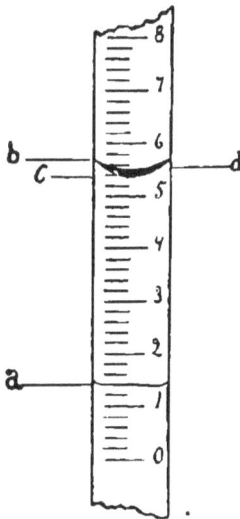
FIG. 5.

pair of dividers, one point of which is placed at the
bottom and the other at the upper limit of the fat.
The dividers are then removed, and one point being
placed at the o mark of the scale on the bottle used,
the other will be at the per cent. of fat in the milk
examined.

Sometimes bubbles of air collect at the upper surface of the column of fat and prevent a close reading;
in such cases a few drops of strong alcohol (over 90
per cent.) put into the tube on top of the column of
fat, will cause the bubbles to disappear and give a
sharp line between the fat and alcohol for the reading.
Whenever alcohol is used for this purpose, the reading
should be taken directly after the alcohol is added, as
after it has stood for a time, the alcohol partially
unites with the fat and increases its volume.

19. Testing Skim Milk, Butter Milk and Whey.
As a small amount of fat is usually present in the
above products, you can get more accurate results by
the use of a special test bottle, which contains twice
as much as the ordinary bottle (generally known as
skim milk bottle). In such a bottle twice the usual
amount of milk and acid can be taken, and the column
of fat being doubled, the reading can be taken with
greater accuracy. Less acid is required for whey than
milk.

20. Testing Cream. Cream can be tested with
the regular Babcock test bottle, by dividing one pipette of cream into two bottles and diluting said cream
with the same amount of water, and finish the test
exactly as with milk, and add the fat of the two bot-

tles for the per cent. of fat in the cream. If the cream is quite rich use three bottles, by dividing one pipette of cream into three bottles, diluting it by adding two pipettes of water equally divided among the three bottles, and then proceed with each bottle as in testing milk, and when completed add the fat of the three bottles for the per cent. of fat in cream. Where a delicate scale is available cream may be tested by weighing about five grams in the bottle, and then multiply the reading by 18, and divide by the weight in grams taken, same as in cheese.

21. A Good Gathered Cream Test. Cream may be tested in ordinary bottles by using a pipette having a capacity of 6.04 c. c. which will deliver about six grams of average cream or one third of the weight of the usual sample. When this pipette is used about 12 c. c. water should be added to the cream in the bottle before adding the acid. The usual amount of acid should be taken and the test completed in exactly the same way as with milk. The reading should be multiplied by three to obtain the per cent. of fat in the cream.

TESTING CHEESE.

22. How to Take the Sample. Where the cheese can be cut a narrow wedge reaching from the edge to the center of the cheese will more nearly represent the average composition of the cheese than any other sample. This may be chopped quite fine, with care to avoid the evaporation of water, and the portion for analysis taken from the mixed mass.

When the sample is taken with a cheese tryer, a plug taken perpendicular to the surface, one-third of the distance from the edge to the center of the cheese should more nearly represent the average composition than any other. The plug should either reach entirely through or only half through the cheese. For inspection purposes the rind may be rejected, but for investigations where the absolute quantity of fat in the cheese is required, the rind should be included in the sample. It is well when admissible, to take two or three plugs on different sides of the cheese, and after splitting them lengthwise with a sharp knife, take portions of each for the test.

23. How to Make the Cheese Test. For the estimation of fat in cheese, about five grams should be carefully weighed and transferred as completely as possible to a test bottle. From 12 to 15 c.c. of hot water are then added, and the bottle shaken at intervals, keeping it warm, until the cheese has become softened, and converted into a creamy emulsion. This may be greatly facilitated by the addition of a few drops of strong ammonia to the contents of the bottle. After the contents of the bottle have become cold the usual amount of acid should be added and the bottles shaken until the lumps of cheese have entirely dissolved. The bottles are then placed in the machine and whirled, the test being completed in the same manner as with milk. To obtain the per cent. of fat the reading should be multiplied by 18, and divided by the weight in grams, of cheese taken.

THE COMPOSITE TEST.

24. Potassium Bichromate. The discovery by Mr. J. A. Alen, a Swedish chemist, that potassium bichromate will preserve milk from coagulation and in excellent condition for testing for a long time, offers the most satisfactory solution to this problem yet proposed. This salt, although poisonous, is not so violent a poison as corrosive sublimate, and may be used with comparatively little danger. On account of its bright, orange color it is not likely to be mistaken for any other substance used in the dairy, and the tint which it imparts to milk, without the addition of any other coloring matter, is so marked that there is no danger of milk that has been treated with it being used for domestic purposes.

25. The Potassium Bichromate Thoroughly Tested. The use of potassium bichromate for the preservation of composite samples of milk has been thoroughly tested, with most satisfactory results, by students of the Wisconsin Dairy School during the winter of 1893. Under the direction of Dr. Babcock, and under the immediate charge of myself, samples of milk have been kept in this way in a warm room for more than a month without being coagulated, and determinations of fat in these samples, at frequent intervals, have shown no change in the amount of fat found. In all, 114 composite tests were made by this method. Each of these was made up of either four or six samples of milk, ranging from partly skimmed milk, containing little fat to very rich milk containing more than 6 per cent. of fat.

The samples were kept in a warm room from eight to ten days after the first portion was added, and were, without exception, in good condition when the final test was made. All determinations of fat, both in the single and composite samples were in duplicate, the bottles containing the tests were inspected by myself and a written report given to me each day of the test. The final results are given below:

Average per cent. of fat in all single samples, 3.676.
Average per cent. of fat in all composite samples, 3.654.

Of the 114 trials there were only four in which the difference between the composite test and the average of the single tests exceeded two-tenths per cent., and in all of these the milk was partly churned by too much mixing, making it impossible to obtain a representative sample of the composite. Of the remaining 110 trials, only ten gave differences larger than one-tenth per cent. fat, and in forty trials the composite test agreed exactly with the average per cent. of fat in the single tests. These results are far better than we have obtained by any other method, and I heartily recommend its adoption in factories, as *The composite test.*

N. B.—A series of composite tests just completed by the Dairy students of the term of '94 have given even better results than the previous winter, as they were instructed to handle the samples carefully to prevent a partial churning.

26. How to get the Samples. In making tests on
this plan a pint or quart fruit jar should be provided
for each patron. Into each of the jars should be
placed, at the start, from one-fourth to one-half gram
of powdered potassium bichromate. This need not be
weighed as the amount can vary considerable without
affecting the results. The amount specified is about
one-half as much as would lie upon a cent or as much
as can be taken upon a pen knife blade one inch long.

This will be sufficient to preserve from a pint to a
quart a week. Enough should be used to tint the
whole sample when complete a light straw color,
and it should be perfectly liquid when the final test is
made. If it does not keep perfectly liquid, more bichromate should be used.

Each jar should have upon it the name or number
of the patron to whom it belongs.

27. To measure and keep the Sample. A small
tin cylinder holding from one to two ounces of milk
when filled to the brim makes a convenient measure
for this purpose. Whenever a fresh sample of milk is
placed in the jar it should be mixed with the milk pre-
viously added by giving the jar a rotary motion. The
jars should be tightly closed after each sample of milk
is added and kept in a cool place during the week.
If kept too warm the cream will become so hard that
it cannot be mixed in without danger of churning
which will always lead to low results.

The test of the composite sample is made in exactly
the same way as with fresh milk.

28. Sampling Milk in Factories. There are several good methods to follow:

First.—By stirring the milk with a long handled dipper after it has been poured into the weigh can, and dipping out a small portion from which the test sample is measured.

Second.—By punching a small hole in the bottom of the conductor pipe through which a small portion of the milk continually escapes and is caught in a small vessel placed to receive it.

Third.—By laying a small tube in the bottom of the conductor pipe having it project a foot or more beyond the end and placing a small vessel to receive the portion of milk which runs through the tube.

Fourth.—By using a small tube about three-eighths of an inch in diameter with a slide so arranged at the bottom that it will close by slipping up over the tube and close up, when the operator presses down the tube. This tube should be at least as long as the weigh can is deep. As quick as the milk is poured into the weigh can, the operator inserts the tube and presses slightly, to close the valve at the bottom, and the tube contains a column of milk from top to bottom. He then lifts out the tube and inserts the upper end of it in the composite jar and pours in the milk. This is the ideal way of getting a correct composite sample.

This tube was first brought to notice by M. A. Scovell of the Kentucky Exp. Station.

29. Quick Sampling When Busy. Suppose you have fifty patrons numbered from one to fifty. First.

—In your refrigerator you have fifty jars numbered from one to fifty; one belonging to each patron.

Second.—A rack containing fifty nine-inch test tubes is placed near the weigh can, also numbered from one to fifty. The tubes in the rack being so arranged that by commencing at one end they may be filled continuously, (no skipping about). Into each tube place a very small amount of powdered potassium bichromate, hardly the size of a wheat kernel.

Third.—Your milk sheet must contain the name and number of each patron.

Fourth.—Prepare a tablet having fifty printed numbers for each day in the week.

When the first patron comes (say number forty), take an ounce measure full of his milk and pour it into the first tube in the rack.

Observe his number when you set down the weight of his milk, and quickly put down his number opposite the printed No. 1 in your tablet. When emptying the samples, at your leisure, into the jars in the refrigerator, we observe that in this case tube No. one would be emptied into jar No. forty, and so on with each patron. Say your next patron is No. twenty-six which goes into tube No. two. Tube No. two will then be emptied into jar No. twenty-six. Very quick work can be done as there is no skipping about or hunting for samples.

30. Another Good Composite Test. A very satisfactory composite test may be made in the following manner:

Take a double size bottle (skim milk bottle) for each

patron and measure into this with a 5 c.c. pipette a sample of his milk each day for seven days. The bottle will then contain double the usual test sample, and by adding double the usual amount of acid, the test may be completed as with fresh milk. It is well to shake up the contents of the bottle before adding the acid. But it should be remembered that when a double test sample is used each division on the neck of the test bottle is then only one-tenth per cent. while with a single test sample it reads two-tenths per cent. A composite test for three days can be obtained in the same way with a common bottle by using a pipette containing 5.9 c.c., making the test in just the same way as with fresh milk.

HOW TO DETECT WATERED MILK.

31. A Simple Formula. After milk has stood from two to three hours the lactometer reading may be and generally is from one to two degrees higher than it was on the same milk immediately after it is drawn from the cow, hence it is quite impossible to get a strictly accurate formula.

But the writer's aim is to give a simple formula only approximately accurate, but nevertheless a very valuable guide which may be quickly applied by any intelligent person in a few minutes' time.

32. Directions for Using the Quevenne Lactometer. For convenience in the following explanation we assume that L. R. means Lactometer Reading. T. means Temperature.

Send to some dairy supply house for a Quevenne Lactometer, and a glass tube about two inches in diameter and ten inches high (or a tin cylinder of that size is sometimes used). Take a sample of the milk you wish to test, mix it well, and pour it into the tube to within three inches of the top. Then insert the Lactometer carefully, and pour in enough milk to fill to the top.

Observe the division of the scale which corresponds with the surface of the milk for the lactometer reading.

Find the temperature of the milk, as the correct lactometer reading is only obtained at 60° F. A lactometer with a thermometer attached is best. Where the two instruments are combined the thermometer scale should be above the lactometer scale so that both readings may be taken without removing the lactometer from the milk. If the temperature should not be just 60° the lactometer reading may be corrected by the following rule.

33. Rule for Correcting the Quevenne Lactometer Reading.—Within the range between 50° and 70°.

First.—If the T. is above 60° add one-tenth to the L. R. for every degree it is above 60°.

Second.—If the T. is below 60° subtract one-tenth from the L. R. for every degree it is below 60°.

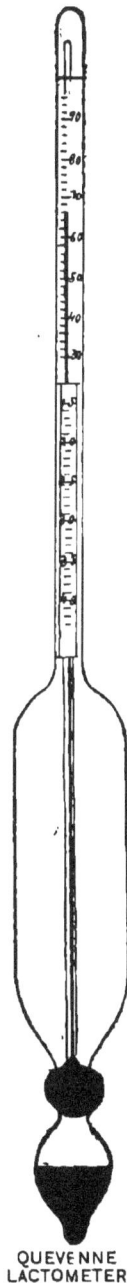

QUEVENNE
LACTOMETER.

34. Examples Under the Above Rule.

L. R. 32, T. 68°.   Correct L. R. 32.8.

L. R. 26, T. 55°.   Correct L. R. 25.5.

35. The Next Step. After you have found the correct L. R. under the above rule, the next step will be to find the per cent. of fat in your sample. Then you are ready to figure for water.

36. About Solids Not Fat. Solids not fat in average milk is about nine per cent., but it may run as low as 8.5 pounds in a hundred pounds of milk. Hence we adopt that as a standard, and for the following reason: Suppose we would adopt 9. as a standard, then all those that have cows giving milk containing less than 9 per cent. solids not fat, could be accused of watering—8.5 is a safe standard.

37. Rule for Finding Solids Not Fat. Multiply the per cent. of fat by .7, add the product to the correct L. R. and divide the sum by 3.8 the quotient will be the solids not fat in your sample.

38. Examples Under the Rule.

1st—Fat 4.   Correct L. R. 32.

4. $\times$ .7 $=$ 2.8 and 32 + 2.8 $=$ 34.8 and

34.8 ÷ 3.8 $=$ 9.16 solids not fat$=$normal milk.

2d—Fat 3.   Correct L. R. 26.

3. $\times$ .7 $=$ 2.1 and 26 + 2.1 $=$ 28.1 and

28.1 ÷ 3.8 $=$ 7.4 solids not fat$=$watered milk.

39. Rule to Find the Amount of Water. Subtract the obtained solids, not fat, from 8.5, multiply the remainder by 100, and divide it by 8.5, the quotient will be the per cent. of water in the sample.

40. Example Under the Rule. We take the above 2d

example: $8.5-7.4=1.1$ and $1.1 \times 100=110$. $110 \div 8.5$
$=13$.   Same as 13 per cent. water in sample.

41. How to Use a Common Lactometer. A common lactometer can be used in place of a Quevenne by observing the following:

Temper your milk to 60° F.

Insert the lactometer and take the reading.   Then multiply the reading by .29, which will reduce it to a Quevenne lactometer reading.   Suppose your reading is 100 and $100 \times .29=29$.   This being what it would read by the Quevenne lactometer.   Suppose, again, the reading is 110 and $110 \times .29 = 31.9 =$ Quevenne lactometer reading.   By observing the above rules a common lactometer could take the place of a Quevenne, but it should be remembered that cheap lactometers are not reliable as a rule.

42. About Testing at the Wisconsin Dairy School. Each student in the Laboratory Section is required to make tests (testing either milk, cream, whey, cheese or butter), and make out a report of his work on a blank furnished by the station.   Composite milk testing, on account of its great importance, has received special attention for the last two winters. Figures obtained by the students under this work may be found in this book.

The following is a blank, showing the work of a student covering one composite test extending over a period of ten days.   In the blank, under Adulterations, we see m96.58 w3.42, which shows that 100 lbs. of the sample contains 96.58 lbs. of milk and 3.42 lbs. of water:

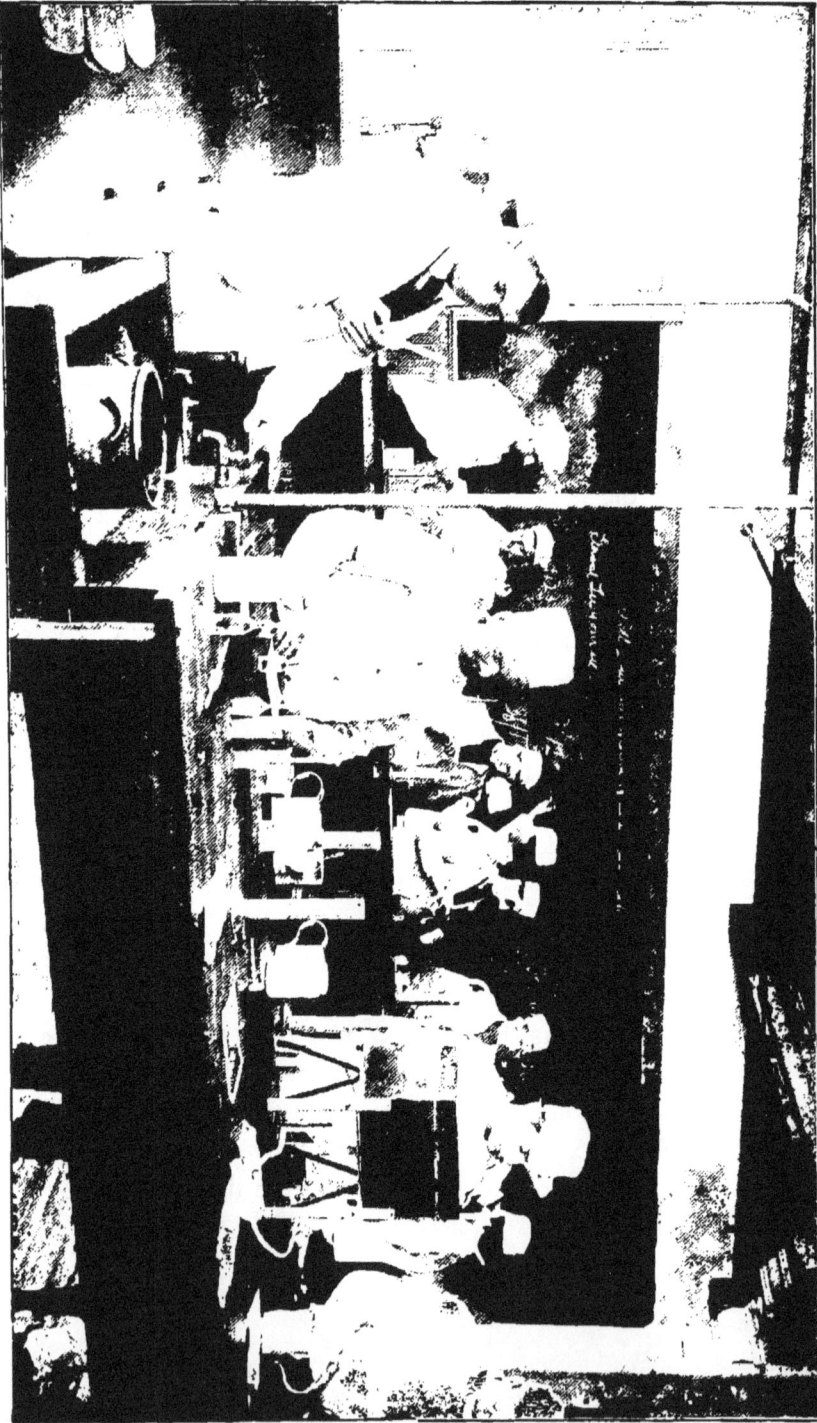

STUDENTS TESTING MILK—WISCONSIN DAIRY SCHOOL.

.

UNIVERSITY OF WISCONSIN—SCHOOL OF DAIRYING.

Report by P. E. WALLINE, No. 73.   DATE, Feb. 14, 1894.

MILK TESTING.

NUMBER OF SAMPLE.	LACTOMETER READING.	TEMPERATURE.	CORRECTED READING AT 60.	PER CENT. OF FAT.	SOLIDS NOT FAT.	ADULTERATION: KIND AND AMT.	FAT BEFORE WATERED.
11	29	58	28.8	3.45	8.21	m96.58 w 3.42	3.57
12	25	56	24.6	3.00	7.02	m82. 6 w17. 4	3.63
70	27	59	26.9	1.90	7.43	m87. 4 w12. 6	2.17 Skim'd.
89	26.5	70	27.5	3.55	7.89	m92. 8 w 7. 2	3.73
82	34	50	33.	2.20	9.09	Skimmed.	....
70	31	67	31.7	2.80	8.86	Skimmed.	....
Total...	......		172.5	16.90	48.50	....	
Average ...	......		28.7	2.81	8.08		
Composite ..28.5		63	28.8	2.80	8.09	....	

N. B.—Students are required to work on a three per cent. fat basis, and an 8.5 per cent. solids, not fat, basis.

HOW TO DIVIDE THE MONEY.

43. The Correct Way.   Let us suppose that there is one composite test taken weekly, and

A bas for the first week 2,046 lbs. milk-test, 3.2 equals fat 65.47.

B   "   "   822   "   "   4.1   "   33.70.

C   "   "   625   "   "   4.6   "   28.75.

A has the second week, 1,820   "   "   3.3   "   60.00.

B   "   "   780   "   "   4.0   "   31.20.

C   "   "   735   "   "   4.2   "   30.45.

A has the third week,	2,244	''	''	3.0	''	67.32.
B	''	''	1,000	''	''	4.2	''	42.00.
C	''	''	650	''	''	4.4	''	28.60.
A has the fourth week,	2,120	''	''	3.1	''	65.72.
B	''	''	962	''	''	4.0	''	38.48.
C	''	''	720	''	''	4.1	''	29.52.

Total, 14,514   Total, 521.21.

A's milk for month, 8,230 lbs. equals 258.51 fat.
B's   ''   ''   3,564 ''   ''   145.38 ''
C's   ''   ''   2,720 ''   ''   117.32 ''

Total, 14.514   Total, 521.21 ''

44. The Butter Sales.

First shipment,	4–40 lb. tubs,	160 lbs.	net amount,	$40.00
Second	''	2–60 ''	''	120 ''	''	''	32.20
Third	''	2–60 ''	''	120 ''	''	''	30.50
Fourth	''	2–60 ''	''	120 ''	''	''	31.60
Home sales,			50 ''	''	''	12.50
A drew			10 ''	''	''	2.50
B drew			6 ''	''	''	1.50

Total, 586 ''   Total, $150.80

Cost of manufacturing 586 lbs. at 4c.   23.44

The patrons' share is,   -   -   -   -   - $127.36

If 521.21 lbs. fat are worth $127.36, 1 lb. of fat is worth, 24.43c.

A's share equals 258.51 × 24.43c. equals $63.16 ⎫
B's   ''   ''   145.38 × 24.43   ''   35.53 ⎬ $127.36
C's   ''   ''   117.32 × 24.43   ''   28.67 ⎭

N. B.—Most secretaries carry over the amount brought about by
the small fractions, to save figuring: If, in the above case, we carry
$2.27 forward to the next month, we would have:

A's share equals 258.51 × .24c. equals $62.04 ⎫ $125.09
B's   ''   ''   145.38 × .24c.   ''   34.89 ⎬ 2.27 carried over.
C's   ''   ''   117.32 × .24c.   ''   28.16 ⎭ ————
                                                    $127.36

45. The Practical Way. Taking the same milk
and the same test we find it as follows: Here we

find the average test by adding the four tests together and dividing by four.

A. 8,230 lbs. milk. Average test for month 3.15 = 259.25 lbs. fat.
B. 3,564 " " " " " " 4.075 = 145.23 " "
C. 2,720 " " " " " " 4.325 = 117.64 " "

Total 14,514 lbs. milk. Total 522.12 " "

N. B.—We find that in this case we have nearly one pound of fat more for the total. By inspection we find that A. has nearly one pound more than in the former statement. While B. has a trifle less and C. has a trifle more.

Of course if cows vary very widely in test and in milk yield from week to week, we would recommend the "correct way." But for general practical purpose the incorrectness of this "practical way" is so slight, and labor saved in figuring so marked that most any one will be justified in using it.

46. A's Butter Statement.

Total No. pounds milk............................	14,514
Total No. pounds butter..........................	586
Average net price for butter......................	25.73 cts
Net receipts for butter...........................	$150.80
For making butter @ 4c...........................	23.44
Net amount due patrons...	127.36
Per pound of butter fat (net to patrons),..........	24.43 cts
General average test	3.59
Your average test...............................	3.141
Your No. of pounds of milk......................	8,230
Your butter fat.................................	258.51 lbs
Your net proceeds...............................	$63.16
10 pounds butter drawn @ 25c...................	2.50

Amount due you............................... $60.66

47. Dividing Cheese Money.

Taking the same figures as in butter:

A's milk for month	8230 pounds.	Fat	258.51	
B's " "	3564 pounds.	"	145.38	
C's " "	2720 pounds.	"	117.32	
Total " "	14514		Total 521.21	

Suppose you get 1,450 pounds cheese selling at ten cents net
    making total amount of money...................... $145.00
Manufacturing of 1,450 @ 1½ cents per pound equals....... 21.75
                              Leaving patrons........................... $123.25

If 521.21 pounds fat are worth $123.25, one pound of fat is worth 23.65 cents.

    A's share equals 258.51 × 23.65c. equals $61.13
    B's   ''   ''   145.38 × 23.65   ''   34.38
    C's   '' ●''   117.32 × 23.65   ''   27.74

             Total,   -   -   -   - $123.25

N. B.—In dividing the money in this case the Secretary might have taken out seventy-seven cents and given it to the patrons the following month. He would then have money $122.48. Fat 521.21. Price for fat 23.5. Saving much labor in figuring.

48. A's Cheese Statement.

Total No. pounds milk...................... 14,514
Total No. pounds cheese......................... 1,450
Average net price per pound.................... 10 cts.
Net receipts for cheese........................ $145.00
Making and selling @ 1½ cents................. 21.75
Net amount due patrons 123.25
Per pound butter fat (net to patrons) ... . 23.65cts.
Your amount of milk .... 8,230. ℔s.
General average test 3.59
Your average test.............................. 3.141
Your butter fat............................... 258.51 ℔s.
Your Net Proceeds............................. $61.13

P. S.—Suppose A had drawn fifty pounds of cheese and 1400 pounds were sold to the buyer. We of course would have charged him the same and would now get our pay by taking it out of the $61.13 leaving him $61.13—5.00 =$56.13.

Before closing these chapters the writer would like to add that recent investigations tend to confirm the fact that satisfactory results can be obtained by making composite tests only every ten days, or even better than that, viz.: semi-monthly. A great advantage under semi-monthly composite tests would be that in making monthly dividends the Secretary would have to deal with no worse fractions than tenths or five one hundredths, either of which is a very easy and simple fraction to handle; the assumption being that tests are read not closer than tenths.

PART III.

USEFUL POINTERS FOR MAKING THE TEST.

49. About Running the Machine.

1. All testing machines should be examined frequently, to see that all the bearings are in order. And especially should those machines having rubber bearing be watched; for loose adjustments lessen the speed of the wheel containing the bottles, and result in a poor test.

2. An ordinary size machine should be run at a speed of about 1,000 revolutions per minute. Small wheels should be run faster, and large wheels slower, ranging from 700 to 1,200 revolutions.

3. The machine should be frequently oiled, and the bearings kept free from dirt and dust.

4. Never hold to the crank of a rubber or belt bearing machine to stop the motion of the wheel suddenly, but let it gradually stop on its own accord. Forcibly stopping a rubber bearing machine may ruin its bearing in a very short time. Remember this, ye operators of rubber bearing machines.

50. About Adding the Acid.

5. Acid of a specific gravity of 1.82 to 1.83 is of the right strength for milk testing; 1.83 is preferable.

6. If the acid is too weak, use a little more than the usual quantity.

7. If your acid is too strong, use a little less than the usual quantity. Do not dilute the acid with water.

8. If your acid is considerably too strong or too weak do not use it at all.

9. Experience has shown that you can greatly improve your test, when your acid is too weak, by warming the milk before adding the acid, but not above 75° F. And when your acid is too strong you can greatly improve your test by cooling the milk before adding the acid.

10. If by circumstances you are compelled to use very weak or very strong acid, observe the advice given in No. 9, and follow it. You will find it a great aid.

11. Always hold the bottle in a slanting position when adding the acid, allowing the acid to flow down along the inside of the bottle to the bottom.

12. Holding the bottle in a straight, upright position and letting the acid drop down onto the milk is a sure way to char your milk and get black spots in your test.

13. Do not let the bottles stand long with the milk and acid unmixed. Do not let the bottles cool off after mixing, but test while hot.

14. Keep your acid bottles tightly corked. Either a glass or rubber stopper should be used. Glass stoppers well fitted to the bottle are the best.

15. If the acid is left open to the air it will absorb moisture from the air and become weak.

16. When the appearance of the fat is of a light

color, and contains a little whitish curdy matter at
the bottom of the fat column, it is a sure sign that the
acid is too weak.

17. When the appearance of the fat is of a dark
colour it is a sign of too strong acid.

51. Distributing the Bottles.

18. In an even numbered bottle tester always put
them in pairs opposite each other.   In a fifteen
bottle tester, to test five bottles, put in one bottle and
skip two places each time until the five are thus dis-
tributed.   To test three bottles, put in one bottle then
skip four places, then put in another bottle, then skip
another four places and then put in your third bottle.
Six bottles can be tested by putting in a pair of bottles, then skip three places, put in another pair and
so on.   For nine bottles put in three each time and
skip two places each time.   For twelve bottles put in
four each time and skip one place.

52. Black Spots.

19. If you are troubled with black spots in your
fat, it generally comes from either of the following:
First — By pouring your acid into your bottle in
such a way that it drops on the top of the milk. (See
No. 11).   Second—By letting your bottle stand unmixed after the acid has been added.   Third—By
having your milk too warm when the acid is added.
If you have a large dark mass in the column of the fat,
give your bottles an extra whirl which generally brings
it to the bottom.

53. White sediment in Fat.

20. If you have a cloudy, white sediment in the lower end of the fat column, it indicates that your acid is too weak, and more acid should be taken. A white sediment test can often be improved. Thus: Cool down the fat in the neck, and then add a little hot water to fill the neck to the figure 10, heat up the bottle quite hot in hot water and whirl lively for a minute.

54. Trouble About the Bulb.

21. In testing thin cream (or when dividing rich cream into two cream bottles), be careful to fill the bottles not above the figure 14. If you fill too high the lower reading will be in the bulb, and no reading can be taken.

If your cream is richer than expected, you can easily fill the second time and give it another short whirl.

55. Miscellaneous Useful Hints.

22. Never make a test without first covering the jacket. The bursting of a bottle in an uncovered machine might cause a serious accident by throwing acid or glass into the operator's face.

23. Experience has shown that a much clearer test can be obtained by filling your bottles only up to the base of the neck, after whirling the first time. Then whirl for a minute and fill to the figure eight and then whirl for another minute. Try it if your test does not show up clear. It works fine.

24. Always use water from condensed steam or rain water in preference to hard water. Soft water tests are much more satisfactory.

25. If for any reason the test can not be made immediately after mixing the milk and acid, set the bottles in hot water to prevent them from cooling.

26. Immerse the bottles, (after the test is completed) in very hot water, completely immersing the fat column in the neck. This will keep the fat in liquid form, and give you a fine reading. If the bottle is allowed to cool before the reading is taken, the fat will be lowered in the neck of the bottle and adhere to the sides, which prevents getting an accurate reading.

27. Especially should the advice given in No. 26 be followed when the atmosphere of the room is cool, or when a large number of tests are made at one time.

28. If your test gives unsatisfactory readings cool down the fat in the neck of the bottle to about 40° F. then immediately immerse in hot water, above the fat column and thereby heat your fat to 150° F. and you will find an improvement in the clearness of your test.

29. Never mix your milk and acid in the test bottle with an up and down motion, but give it a rotary motion and you will have a bottle with a clean neck to start with.

56. Study, 1, 4, 9, 11, 13, 16, 23, 25 and 26, especially under Useful Pointers for making the test.

PART I.

APPLICATION OF THE TEST.

PART II.

THE GLASSWARE AND MACHINERY OF THE BABCOCK TEST.

Making the Test.

PART III.

USEFUL POINTERS FOR MAKING THE TEST.

www.ingramcontent.com/pod-product-compliance
Lightning Source LLC
Chambersburg PA
CBHW022024190326
41519CB00010B/1587